The Roman Engineers

L. A. and J. A. Hamey

Published in cooperation with Cambridge University Press
Lerner Publications Company, Minneapolis

Editors' Note: In preparing this edition of *The Cambridge Topic Books* for publication, the editors have made only a few minor changes in the original material. In some isolated cases, British spelling and usage were altered in order to avoid possible confusion for our readers. Whenever necessary, information was added to clarify references to people, places, and events in British history. An index was also provided in each volume.

LIBRARY OF CONGRESS CATALOGING IN PUBLICATION DATA

Hamey, L. A., 1918-
  The Roman engineers.

  (A Cambridge topic book)
  Includes index.
  Summary: Discusses the methods and achievements of Roman builders who were responsible for such projects as aqueducts, roads, and bridges.
    1. Civil engineering—Rome—History—Juvenile literature. 2. Rome—Antiquities—Juvenile literature. [1. Civil engineering—Rome. 2. Rome—Antiquities] I. Hamey, J. A., 1956- II. Title.
TA16.H35 1982   624'.0937   81-13746
ISBN 0-8225-1227-0 (lib. bdg.)   AACR2

This edition first published 1982 by Lerner Publications Company by permission of Cambridge University Press.

Original edition copyright © 1981 by Cambridge University Press as part of *The Cambridge Introduction to the History of Mankind: Topic Book.*

International Standard Book Number: 0-8225-1227-0
Library of Congress Catalog Card Number: 81-13746

Manufactured in the United States of America

This edition is available exclusively from:
Lerner Publications Company, 241 First Avenue North, Minneapolis, Minnesota 55401

1 2 3 4 5 6 7 8 9 10 86 85 84 83 82

Contents

The Romans took the semi-circular arch and made it the most characteristic feature of their architecture. This splendid example is the triumphal arch of the emperor Trajan, built about the middle of the second century AD.

cover: *The picture on the front is an artist's reconstruction showing the building of the Roman bridge over the River Tagus, near Alcántara, Spain, about AD 106 (see page 36). The photograph on the back shows how well the engineering skill incorporated in the design of the bridge and the high standard of workmanship have enabled the bridge to withstand the test of years.*

1 Engineers in Roman society

Most people today have one particular job in society: a person is an engineer, architect, soldier, or bricklayer, and so on. Many professions are themselves sub-divided and these specialities are followed as quite separate careers. Thus an engineer will be either a civil engineer or an electrical engineer, a mechanical engineer or some other type of specialist engineer: each will belong to a specialist technical institute, undertake quite different work, and have only a limited knowledge of the others' trades.

All this was much less true in the ancient world. An engineer would deal with both fixed works and mechanical devices. Today a distinction is also made between civil engineers and architects. Architects are concerned with the arrangement, appearance, and finish of all kinds of building, whilst civil engineers become involved in buildings only when structural problems have to be solved or special foundations are needed.

In the Roman world all these activities would have been the concern of the same person. Thus when Pliny, who held important offices under the Emperor Trajan towards the end of the first century AD, needed advice on a construction project, he asked Trajan to send him 'a surveyor or a civil engineer'. By this he meant that he wanted someone experienced in construction work of any kind.

The experience and training of such men, backed by the traditional skill of the craftsmen, had to provide the information which is today incorporated into the detailed drawings produced before any project is begun. Yet it is clear that work on a complex structure such as the vast Pont du Gard in France could not begin until the position of each pier, the level of each tier of arches, the radii of the arches, and even the size and shape of each arch stone had been established. Perhaps some detailed drawings were made; we do not know, since none has survived.

Today the civil engineer is further distinguished from the military engineer; their work is similar, but military structures are often temporary and their construction techniques are governed by the need for speed and mobility. This clear distinction between the civil and military was blurred in early Republican Rome, as in most city-states. In wartime the farmer became a citizen soldier; there was no professional standing army until the reforms of Gaius Marius, consul in 107 BC.

Under the Empire, which began after the Battle of Actium in 31 BC, the standing army undertook many engineering projects. At first these projects were military. In March AD 107 a young soldier in Sinai called Julius Apollinaris wrote to his mother telling her about his promotion, due to which, he said, he had escaped the stone-cutting on which his comrades were engaged. This stone was probably destined for road-building in the newly conquered province of Arabia.

By the beginning of the second century AD the Roman army had absorbed many of the Empire's best engineers and surveyors. It included a corps of skilled craftsmen, known as *fabri*, under the command of an officer called the *praefectus fabrum*. Public works, especially in the provinces, came to depend heavily on military engineering skill and practical expertise. This tendency accelerated under Hadrian (AD 117–38), as the legions were moved about less and they were able to

The plate of an architect who was associated with the reconstruction of the large theatre in Pompeii.

The imperial authorities thought that one way to civilise barbarous new subjects like Britons was to encourage them to build Roman-style towns as their tribal centres, and lent engineers to plan and direct the work. The Regnenses, who lived in what is now Sussex, built Noviomagus Regnensium as their capital, and the plan shows the outline of the town, parts of which still stand or have been discovered by excavation. The air view of Chichester, as the town is now called, shows the same walls and street plan, and a market cross and cathedral where the main streets cross.

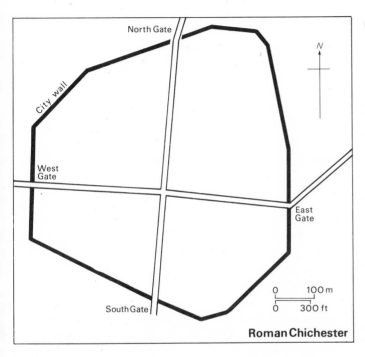

Roman Chichester

give more help within the provinces where they were permanently stationed. From the time of Marcus Aurelius (AD 161-80), military involvement in civilian projects increased steadily until the author of the *Historia Augusta,* writing of Egypt in the reign of Probus (AD 276–82), could say: 'He [Probus] built bridges, temples, colonnades, and public buildings with army labour; opened up many river mouths, drained many marshes and created in their stead fields of grain and farms.'

These changes from century to century remind us that we must always be sure which period we are discussing. According to tradition Rome was founded by Romulus in 753 BC;

Romulus Augustulus, the last Emperor in Rome, was overthrown in AD 478. So the history of Roman civilisation spans twelve hundred years, though we know very little about the early period.

We must remember, too, that building methods, like everything else, would have changed over this long period. On the other hand, the Romans were not a people to change their methods quickly; having found a good technique, they did not automatically look for an easier and cheaper method. The history of Roman engineering is therefore one of slow improvement rather than of sudden change.

Organising public building

In the days when the Roman Republic was beginning to dominate central Italy two needs were pressing: first, means to supply the essentials of life to a city already huge by ancient standards and, second, aids to secure the conquered territories in Italy. The Aqua Appia, an aqueduct bringing water to Rome from Tusculum, and the Via Appia, a road connecting Rome with Capua, were both built during the fourth and third centuries BC to serve those needs.

The inspiration for such projects would come from within the aristocracy or the Senate, though the Populus Romanus (the entire Roman people) might be asked to support the proposals, voting in Assemblies.

A building project was administered either by a Senatorial Commission or by a single man holding the elected office of *censor*, such as Appius Claudius the Blind, censor in 312 BC, after whom the Aqua Appia and the Via Appia were both named. The *aediles*, who were elected magistrates much lower in rank, were responsible for maintaining public works. Both officers were politicians and not paid public servants.

In the late Republic so many works had been built that responsibility for certain categories was taken away from the traditional magistrates and given to men holding offices especially created. Among the new posts were a Curator Aquarum (Director of Water Supplies) and a Curator Viarum (Director of Highways); Julius Caesar was Curator Viae Appiae for a time.

Under the Empire a very different system operated since all real power had become centred on the Emperor. Augustus set the precedent in 20 BC by making himself Curator Viarum. Although a major work of construction might be suggested by a provincial governor, the Senate, or a local city-council, in almost every case the Emperor's approval would be sought.

When Pliny was Governor of Bithynia around AD 112, he asked the Emperor Trajan's approval for a canal. Their correspondence demonstrates clearly the detailed and practical approach of both men, which is particularly impressive in Trajan's case, since as Emperor he must have had a thousand more important matters of state on his mind.

'Pliny to the Emperor Trajan

'It seems to me, as I survey the sublimity of your station and ambition, wholly appropriate to bring to your notice works which are worthy of your immortal name no less than your glory; and likely to possess magnificence with utility.

'There is a sizeable lake in the area of Nicomedia. Across this, marble, farm produce, wood and timber are conveyed in ships with little effort or expense right to the road, from which, with great effort and even greater expense, vehicles take them to the sea. To connect the lake with the sea would require a lot of labour; there is, however, no shortage in the area. A good supply of men is available in the countryside and still more in the town; it seems certain that everyone would contribute most willingly towards a scheme of universal benefit.

'It remains for you, if you think fit, to send a surveyor or an engineer to determine by a careful survey whether the lake is higher than the sea; the local experts maintain that it is, by forty cubits [probably about twenty metres]. In the same place I have myself come upon a canal dug by a former king of Bithynia, but whether intended to drain the surrounding fields or to join the lake with the sea is not clear, since it is unfinished. It is also uncertain whether the work was abandoned because the king died or because he had despaired of completing it. This, however, itself encourages and spurs on my desire – you will bear with my ambition for your greater glory – that you should accomplish what kings have merely begun.'

'Trajan to Pliny

'This lake of yours intrigues me, and I should like to see it connected to the sea, but there must be a thorough and accurate reconnaissance of the source and quantity of water flowing into it, otherwise, once given an outlet, it may all empty straight into the sea. You may apply to Calpurnius Macer for a surveyor, and I myself will send you someone experienced in this type of work.'

(Calpurnius Macer was the nearest army commander who could have had a surveyor on his staff.)

Under the Empire the *curatores*, too, were appointed by the Emperor. He would have looked for men of tried ability willing to devote their whole time to the task; one such man was Sextus Julius Frontinus (page 17), Curator Aquarum under Trajan. His extensive knowledge enabled him to write an authoritative textbook on water supply. Throughout Roman history, however, the full-time engineer was only an adviser: a project was always directed by a politician or administrator.

7

2 Aqueducts

Finding water

Human settlements must always be near a source of fresh water, whether a river or a spring. While Rome was just a small state within Latium, its source was the River Tiber. By the late fourth century BC, when the Romans were fighting the second Samnite War, an alternative source of water was urgently needed. Perhaps this was because the water supply from the Tiber was not reliable enough for the expanding population of Rome, or perhaps it was because a single source of water could easily be poisoned by an enemy. Consequently the Romans began building their first aqueduct, the Aqua Appia, in 312 BC. Most of the later aqueducts were built less urgently to satisfy an amazing and ever-increasing demand for cold, clear water. To understand the builders' task, we must look first at the geography of Rome and its hinterland.

The countryside around Rome, known as the Campagna, is surrounded by hills and mountains. Rain and melting snow from this high ground feed many rivers, among them the Tiber and its tributary the Anio.

Some of this water, however, percolates through the soil and flows over the surface of the hard rock beneath, emerging from the ground lower down as springs. The many springs in the hills around the Campagna captured the attention of the Romans, and they decided to channel some of this pure water into the city.

One might have to dig to find a good source; where should one look? Vitruvius, a retired military engineer, who wrote ten books on architecture and engineering in the reign of Augustus, advised the Romans that

'Finding water is easy if there are open, running springs. If not, we must search underground. Just before sunrise, lie face downwards on the ground, resting your chin in your hands. Take a look over the countryside; where you see vapour curling up from the ground you will find water

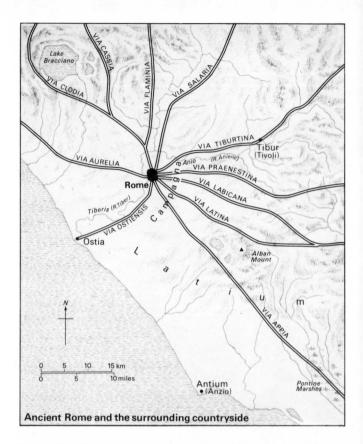

Ancient Rome and the surrounding countryside

if you dig. You must also study the soil. In clay and fine gravel, water will be poor in quality and taste; in coarse gravel it will be sweeter and more reliable; in tufa and lava it will be plentiful and good. If the spring is free-running and open, look at the people dependent on it: if they are strong, have fresh complexions and clear eyes, then the water is good.'

Vitruvius also described another method of checking for underground water in which a bronze bowl was left in a pit overnight then examined in the morning to see if moisture had condensed in it.

Problems and solutions

The Romans' word *aquaeductus*, a conductor of water, tells us what an aqueduct is. It is not a bridge carrying water, though

Specus roofs

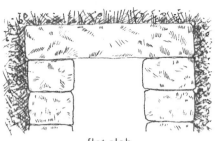

flat slab

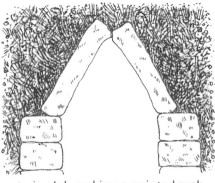

twin slab making a pointed arch

half-round arch

there can be bridges in it; it is any man-made channel carrying water. The Romans preferred the water in their aqueducts to flow downhill, under gravity, so that there was no need to use pressure to make it flow.

Although Vitruvius describes three kinds of aqueductus—masonry conduits, lead pipes, and earthenware pipes—we shall look just at masonry conduits, since these were most common.

The heart of a masonry aqueduct is the *specus*, or water-channel, itself. It was about the same size as a doorway in a modern house. Whether the specus was built under, on, or above the ground, it would have had a stone floor, stone walls, and a stone roof, except where, as in some later aqueducts, concrete was used instead of stone. Occasionally, too, the specus was cut through solid rock.

Vitruvius said that the channel 'should be arched over to shield the water against the sun', but there was another reason for covering the channel in early Rome: a roof of stone slabs made it harder for an enemy to cut off or poison the water. For this reason the earlier aqueducts were built low, and ran as far as possible underground.

Roofs were of three main types: flat slabs, twin slabs leaning against each other to form a pointed arch, and half-round arches. The pointed arch is worth special mention.

The Romans almost certainly learnt about the arch from the Etruscans, who had carried out civil engineering works in this part of Italy long before Rome became a great city. At its simplest, an arched roof was two roughly hewn slabs leaning against each other, the top and bottom inside corners of each

Part of the specus of the Anio Novus aqueduct built about AD 52. The concrete walls of the specus here were faced with brick on the outside and lined with mortar on the inside.

being broken off to make it a little more stable. Gradually, more accurate cutting at the apex improved the joint, which was sometimes strengthened with a crude type of mortar. From such simple beginnings the famous rounded Roman arches eventually developed.

When built in ordinary soil or gravel the aqueduct would be constructed in a trench; where it traversed a steep hillside in soil or gravel there was, as we shall see, endless trouble once the aqueduct was in use. Where, by contrast, there was rock close to the surface, the specus would be cut directly into it unless the route would have required too deep a trench; then it was often easier to cut a complete rock tunnel.

9

If a valley had to be crossed at right-angles, a wall or bridge would be built to carry the channel; earlier builders fought shy of such works and would loop their aqueducts along the sides of each valley.

Within sight of Rome, as the channel of the aqueduct emerged from the higher ground, it was carried on a long, continuous row of arches until it reached the city wall. It is these viaducts which most people think of as 'the aqueducts of ancient Rome'. They were magnificent indeed. The Aqua Claudia, for example, swept into the city on a viaduct of over a thousand arches. The sight of many such viaducts from the southern gates must have filled the citizens of Rome with pride.

Supplies from the Anio

The elder Pliny, writing after two aqueducts—the Aqua Anio Vetus and Aqua Marcia—had been carrying the Anio water to Rome for many years, described its water as 'the most famous in the whole world for coldness and wholesomeness'. We are now going to look at the four aqueducts which were eventually built from the Anio valley to Rome, and which were the most important ever to serve the city.

The first of these four, the Aqua Anio Vetus, was not in fact the first aqueduct built to Rome: this was the Aqua Appia (page 8), whose source lay to the east of Rome in the area

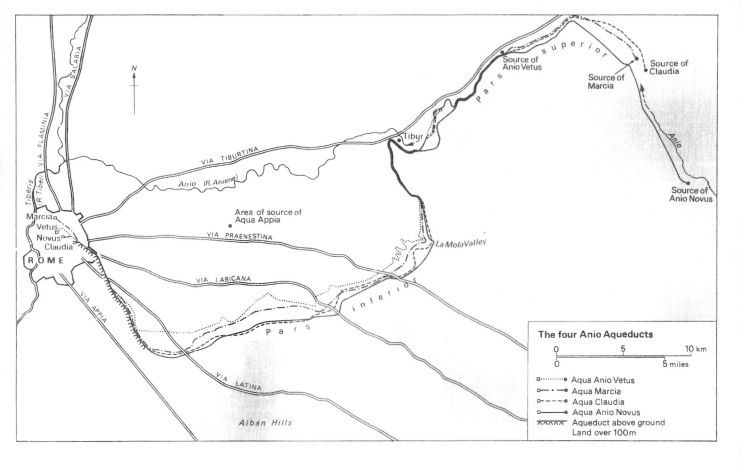

The four Anio Aqueducts

0	5	10 km
0		5 miles

- Aqua Anio Vetus
- Aqua Marcia
- Aqua Claudia
- Aqua Anio Novus
- Aqueduct above ground
  Land over 100m

marked on the map by a dot. It was a mere seventeen kilometres (10½ miles) in length; only forty years after its construction the Romans were building an aqueduct sixty-four kilometres (40 miles) long!

The shortest route from the upper Anio valley to the city lay straight down the valley, but such a route would have given too steep a fall to the upper section of the aqueduct, known to the Romans as the *pars superior*, and too gentle a slope to the lower section, the *pars inferior*. At the junction of these two sections the pressure of water from the steeper pars superior would have burst the channel or caused it to overflow.

This problem was solved by taking the pars superior down the Anio valley at a more gradual slope than the river as far

as Tibur (modern Tivoli), then looping it southwards towards the Alban Hills to enter Rome from the south-east, parallel to the Via Appia.

There are two great difficulties in trying to describe how these aqueducts were built. The first is that practically nothing was written about them; although a good deal was written about the army, the lives of prominent Romans, or even farming, the practical details of building did not interest any writer whose books have survived. True, Vitruvius wrote his textbook, *De Architectura*, but even he did not think it worthwhile to describe the everyday tasks of building-workers. The second difficulty arises because aqueducts were always being strengthened or repaired, and so the original work has often disappeared or been covered up.

NAME AND DATE		LENGTH				HEIGHT ABOVE SEA LEVEL				FALL		
		Total		*Above ground*		*Source*		*Rome*				
		km	miles	km	miles	m	ft	m	ft	m	ft	*proportion*
Anio Vetus	272 BC	64	40	0.3	0.2	264	866	44	144	220	722	1/291
Marcia	144 BC	92	57	11	6.8	313	1027	53	174	260	853	1/354
Claudia	AD 47	69	43	15	9.3	314	1030	66	216	248	814	1/280
Anio Novus	AD 52	87	54	14	8.7	492	1614	64	210	428	1403	1/203

The table shows the four aqueducts in order of construction and compares the length, portion above ground, and fall of each. Note how each aqueduct was started further up the Anio valley than its predecessor so that the water would reach the higher areas of the city. This is well illustrated by the cross-section of a hillside near Tivoli, where the remains of three of these aqueducts can be seen one above the other.

Great lengths of the aqueducts had to be constructed along hillsides. Here, because of the sloping ground, the downhill wall received less support from the earth than the uphill, and this side of the channel would soon develop leaks; these would soften the ground, causing the wall to slip further. Unless repaired or strengthened, the channel would eventually collapse altogether. A pointed roof (page 13), which in any case tends to push the walls outwards, would add to this problem. The sections carried on viaducts, too, needed frequent repair. The writers Juvenal and Martial, who were writing shortly before AD 100, complained of water dripping from the Porta Capena, a gate in the city wall which carried part of the Aqua Marcia over the Via Appia.

We must draw on our imagination, then, if we are to watch the construction of one of these four aqueducts. We have chosen the Anio Vetus, the first to bring Anio water to Rome. (The word *vetus,* meaning old, was added when the new Anio was built 300 years later.)

When Manius Curius Dentatus undertook (as censor) the building of the Aqua Anio, the long Italian wars were ove In 290 BC, at the end of the Third Samnite War, he ha celebrated a triumph over the Samnites, and in 272 BC th last truly independent state in Italy, Tarentum, had su rendered to his forces. In the meantime the Via Appia ha

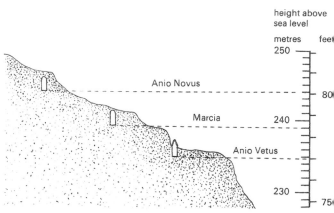

Section drawing of three aqueducts as they emerge from a hillside near Tívoli.

Section diagram of an aqueduct channel with a pointed roof, built along a hillside. There is much less earth support to the downhill wall.

been extended through the heart of Samnium to Venusia, and the Via Flaminia had begun to creep northwards. Rome, Mistress of Italy, was in an expansive mood, and around 270 BC began to build the Aqua Anio.

The Senate's first consideration would be an estimate of the cost. To assess this the Senatorial Commission which would eventually control the project had to decide which springs to tap, and would then instruct a surveyor to set out the approximate route.

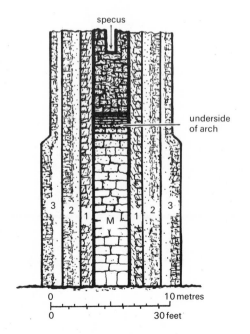

Anio Novus

Claudia

above: The Praenestine Gate in the city wall at Rome was used to carry the Claudia and the Anio Novus above. (The building on the right, immediately outside the city wall, is the tomb of Eurysaces the baker; the circular openings represent ovens.)

left: Section diagram of a tall, narrow bridge carrying the Aqua Marcia across the Mola Valley. It had to be strengthened by Augustus (31 BC to AD 14), Titus (AD 79–81), and Hadrian (AD 117–36).

M – original viaduct of limestone masonry
1 – first strengthening: concrete faced with stone
2 – second strengthening: concrete faced with brick
3 – buttresses of concrete faced with brick

13

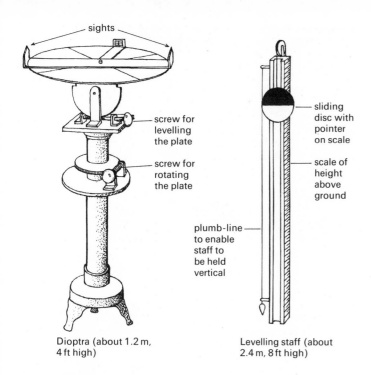

sights

screw for
levelling
the plate

screw for
rotating
the plate

plumb-line
to enable
staff to
be held
vertical

Dioptra (about 1.2 m,
4 ft high)

sliding
disc with
pointer
on scale

scale of
height
above
ground

Levelling staff (about
2.4 m, 8 ft high)

left: *The dioptra. When the surveyor was taking levels he
would first adjust the horizontal plate with a small water level;
he would then sight on the staff held by an assistant and signal
to him to move the target disc up or down until it lay on the
horizontal line of sight.*

below: *In the diagram the aqueduct has its source at A and will
emerge from underground at B. Before the surveyor can mark
out the length XY for men to start work, he has to obtain both
the distance between A and B and their difference in height.*

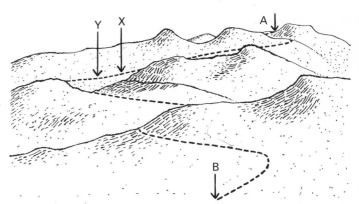

Surveying the route

The task of accurately setting out an efficient water-channel
over sixty-four kilometres (40 miles; two days' journey on
foot) must have been daunting. The approximate route set
out by the *librator* (surveyor) would follow a fairly even,
gentle slope between the source and the city. The librator
would mark the route with wooden stakes. Then he would
have to find out the difference in height between the source
and the city. The Romans had no ready-made maps and had
to measure carefully for each task. Before the workmen could
begin on any one section of the aqueduct, the librator had to
calculate its overall fall and establish the height of each end
of the section.

The difference in height (as we now know from our maps)
between the source and outfall of the Aqua Anio Vetus of
220 metres (722 feet), and the length of about 64,000 metres
(40 miles) give an overall fall of 1 in 291. A Roman surveyor's
levelling instrument, the *dioptra*, had a very limited sighting
distance, so, in order to take the measurements, the librator

would have to set up his instrument several hundred times over
the whole distance. All the separate differences in level would
be recorded on his wax tablet and added together to give the
total difference in the height of the land between the source and
the city. Once the librator had measured the approximate
length of the aqueduct, and the difference in height between
the source and the city, he could calculate the overall fall and
then set to work on marking out the final route.

Establishing the final alignment of the aqueduct alongside
the approximate route was complicated and laborious,
largely a matter of trial and error. It must lie as close as
possible to the outline route, avoid sharp bends, keep to the
overall fall of 1 in 291, and never rise above ground except in
the final entry to Rome.

When, after weeks of painstaking work, the full length of
the correct line had been established, stout wooden posts
would be set up at regular intervals on either side of the
planned path of the aqueduct to replace the stakes of the rough
alignment.

above: *When the surveyor has taken a backsight (Y) and a foresight (X), he can calculate the difference in ground levels over this short distance. The process must be repeated over the whole route of the aqueduct to obtain the overall fall.*

below: *The chorobates, a long narrow table, was used to check whether work was level or had been correctly sloped. A plumb-bob at each corner and marks on the leg bracings showed level. Water could be used in the groove down the centre instead.*

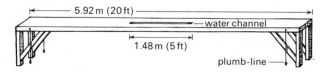

5.92 m (20 ft) — water channel

1.48 m (5 ft)

plumb-line

Building the aqueduct

The men working on an aqueduct would have to sleep and eat in camps spaced along the route; since transport was slow, these camps would be isolated and self-sufficient. Workmen would come from a variety of places. Some, no doubt, would be slaves, especially those employed in unpleasant jobs like tunnelling and stone-breaking; others would be labourers hired for the day from the small towns of Latium and others would be unemployed workers from Rome.

Work would begin simultaneously at a number of points on the aqueduct; all along the route gangs would be taking off the top-soil, levelling bumps and filling hollows, then ramming stone into soft places to make a good temporary road for the pack-animals and heavy carts.

The building-stone would not be transported any further than was absolutely necessary; fortunately, the Anio valley contains good rock, so quarries could be set up near the route of the aqueduct. Every block of stone would have to come some way, however, in carts, panniers or on skids.

For most of the gangs the preliminary work would be straightforward trenching in soft ground; the sides of each section of the trench would have to be temporarily secured with timber props. For the less fortunate workmen part or all of their trench would have to be broken out of rock. The un-luckiest of all would be allotted to the tunnelling. They had first to sink a *puteus*, or shaft, every seventy-one metres (233 feet) or so; then, with just enough room for one man to work at the rock face, they would tunnel forward, passing back the hewn stone in baskets to be hauled up the shaft. Without free-flowing air, the atmosphere would soon become foul from the oil-lamps and the breathing of the workmen.

At all stages the libratores would check the progress; once the channel had been roughed out, a *chorobates* would be lowered into the open trench to check the slope. This instru-ment was too big to lower down a shaft, so a water-level would be used instead in the tunnels.

At the same time other workmen, probably slaves, would be bringing up stone, roughly hewn at the quarry into blocks about half a metre square by one to one and a half metres long (20 in. square by 40–60 in.). Each block would be carefully dressed on the site by masons; they paid particular attention to the horizontal or bedding face and to internal joints which must fit tight without mortar between them, though the channel would be lined with mortar to ensure that it stayed clean inside.

Most of the roof on the Vetus was a pointed arch. Imagine that we are watching part of the roof being placed in position. The walls of the specus are finished and their tops can be seen below us in the open trench. On each side of the trench a gang of four men is slowly lowering a heavy slab. When the bottom edge of each slab is resting on the wall, the slabs are allowed to fall gently together until their tops meet. The ropes will be kept taut until the earth has been rammed down behind both slabs to prevent either from pushing the other upright.

When the ropes have been cast off, the whole specus is buried and the soil rammed hard. Any surplus is spread so that the ground will appear undisturbed once the vegetation returns; it is only thirteen years since a battle was fought within eighty kilometres (50 miles) of Rome.

left: *Building an aqueduct. The artist has tried to reconstruct as many of the builders' tasks as possible, though it is not likely that so many would be going on so close together at the same time. This section is being cut in very soft ground so the trench sides have to be supported by timber which will be buried once the aqueduct roof is complete.*

right: *The aqueducts had to provide for all the normal needs of a very large population and also supply the baths, large leisure centres which must have used great quantities of water. In the early years of the city drainage had been provided naturally, by the slopes and streams, but as the city grew this was converted into a system of sewers.*

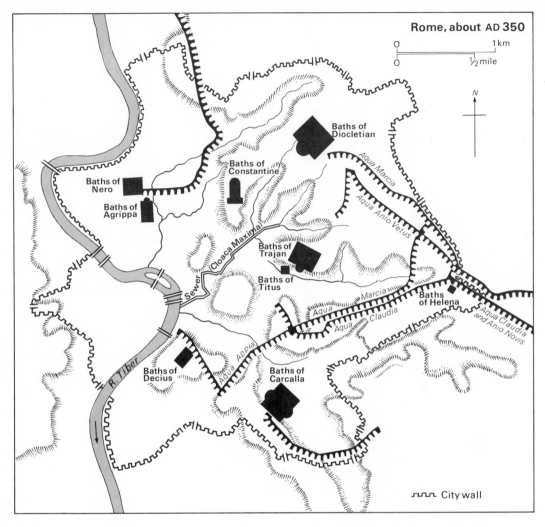

Rome, about AD 350

0 ——————— 1 km
0 ——————— ½ mile

N

Baths of Diocletian

Baths of Constantine

Baths of Nero

Baths of Agrippa

Cloaca Maxima

Sewer

Baths of Trajan

Baths of Titus

Aqua Marcia

Aqua Anio Vetus

Aqua Marcia

Aqua Claudia

Baths of Helena

Aqua Claudia and Anio Novis

R. Tiber

Aqua Appia

Baths of Decius

Baths of Carcalla

ᒧᒪᒧᒪ City wall

Once complete the aqueduct began its long working life. It needed to be inspected and maintained; it was often repaired or improved.

During the Empire, when the wars in Italy were ended and there was no longer any need to keep the route of the aqueduct secret from the enemy, aqueducts were often marked with milestones, called *cippi*. These were particularly useful for locating specific points on the aqueduct. If, for example,

leaks had developed which needed repair, the position of the leaks could be reported with reference to the nearest cippus.

Frontinus, Curator Aquarum

No account of aqueducts can omit Sextus Julius Frontinus, thrice consul, a Governor of Britain, and Curator Aquarum under Trajan at the end of the first century AD. This distinguished Roman prepared his *De Aquis* as a record of his

right: *Cloaca Maxima, the Great Drain, oldest and biggest of the sewers of ancient Rome, where it reaches the River Tiber.*

below: *In some places the Romans discovered that the water had extra medicinal properties. Roman Aquae Sulis has become the English city of Bath; people still go there to 'take the waters' and, as the photograph shows, the Roman baths are still in a good state, even to the lead piping laid in the floor.*

inquiries into every aspect of Rome's water-supply. He must have turned the Water Department upside down!

By this time there were nine aqueducts serving Rome. Frontinus had maps made of them showing the position and size of all tunnels, bridges and structures. He also had a record made for each aqueduct of the amount of water entering it at source and being discharged in the city. Archaeologists have used Frontinus' figures to estimate how much water flowed to Rome. For example, the figure he gives for the source of the Aqua Marcia is thought to correspond to a flow of just over one million litres (220,000 gallons) of water an hour.

You may think that the amount of water entering the aqueduct should be the same as that discharged in the city, but Frontinus was well aware that he would find considerable differences. It was common to steal water by secretly connecting a pipe to the aqueduct, even though, according to Livy, 'vigorous measures for suppressing this abuse' had been introduced by Cato the Censor two centuries earlier.

Rome always attached great importance to safeguarding water supplies to street fountains, from which the ordinary people drew their water, and to the public baths. Both of these had priority in times of drought over the private connections to great houses and commercial premises. These private and commercial consumers paid for their water; their connecting water pipes would be limited to a certain size, which would be stamped on the pipes and recorded to ensure that additional or larger pipes were not installed in secret. In spite of this, illegal connections were made.

3 Roads

Instruments of Empire

The steady growth of the Roman system of trunk roads to a peak of over 90,000 kilometres (56,000 miles) followed a consistent pattern. All major roads were built initially by or for the army, and so frequently ran outside the Roman world into the hostile lands beyond. At the same time the construction of roads and replacement of old tracks improved communications within the empire for both the army and the government, and, in time, for commerce and the general public.

Roman roads, although built with military needs in mind, were of immense benefit to civil life. A few roads were purely commercial; one was the Via Salaria, the Salt Road, ancient beyond the memory of man, along which salt came to the Sabines from the Adriatic Sea. The Sabines then carried the salt westwards to trade with the Romans.

Before the Romans, other ancient peoples had built outstanding roads. Ancient writers tell us of some which existed long before the Via Appia; they were generally royal roads, built for the convenience of rulers. Herodotus (c. 485–425 BC), a Greek of Halicarnassus in Asia Minor, describes the main road of the Persian Empire.

'All along the road there are royal posts with excellent inns. It runs entirely through inhabited country and is safe to travel by. If the measurement of the Royal Road in parasangs [a Persian measure roughly equal to $5\frac{1}{4}$ km] is accurate . . . the distance from Sardis [in modern Turkey] to the Palace of Memnon [in Iran] will be 450 parasangs [2,400 km] . . . and the journey will take just ninety days.'

Roman roads are so famous that we are apt to regard as typical, achievements which were outstanding even by Roman standards. The most famous Roman road is probably the first, the Via Appia, built in 312 BC. Procopius, a Greek of Caesarea

The Appian Way, near Rome. Although worn by 2,300 years of traffic so that joints no longer fit tightly, the stones of the Via Appia still show clearly how the 'Queen of Roads' earned its name.

in Palestine, writing eight hundred years later, described the work as so perfect that it seemed as if Nature had performed it rather than man, for the very joints were hardly perceptible. Conceived and built according to a monumental design, the Via Appia still survives today in its original form over great lengths. However it is not a typical example of a Roman road. The great genius of the Romans, in road-building as in everything else, lay in their ability to adapt their methods to their requirements and local resources. Even today Italian peasants call a good road a *via appia*!

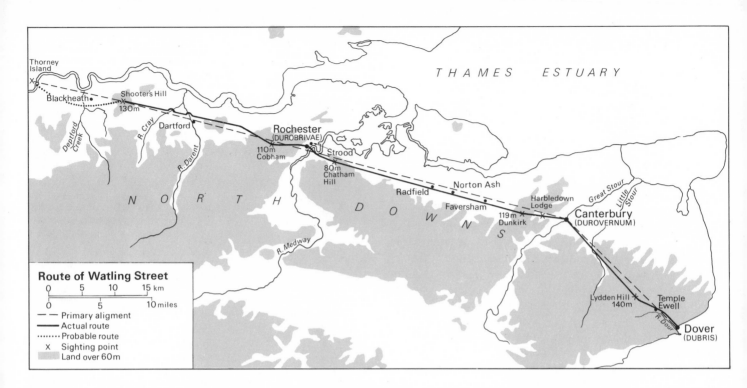

Setting out the route

Roman roads are particularly famous for their straightness, but we must not imagine them as lines ruled on a map. In the first place, the Romans had neither true-to-scale maps nor compasses, and their surveying instruments would be much less accurate than modern theodolites, which depend on the optical lens. We can be sure, however, that the army officers had a remarkable mental grasp of the geography of the areas in which they operated.

Roman roads follow very direct routes and run quite straight over considerable distances, but if you look at a map, you will often find that a Roman road, though following a direct route, zig-zags slightly in a number of straight sections, each a few kilometres long.

Look, for example, at Watling Street, the first and most important road in Roman Britain. This ran from Dover (Dubris) to Canterbury (Durovernum), then turned west through Rochester (Durobrivae) to Thorney Island, where it

originally crossed the Thames to turn north on the line of the present Edgware Road for St Albans (Verulamium) and, finally, Chester (Deva). Later a crossing was made at London Bridge. This has always been a vital route; from Dover to Dartford the modern road, the A2, follows the old road exactly. West of Dartford the modern trunk road diverges, but Watling Street can still be followed along the A296, A226, and A207 until it is lost on Blackheath. For two thousand years, until the M2 was built from Strood to Faversham, this road carried much of the traffic from the Continent into England.

No-one has explained satisfactorily how the Romans set out the route of their roads. Looking at Watling Street, we can see that when it leaves Canterbury it is already pointing directly at Westminster Abbey, which stands where Thorney Island once formed a natural stepping-stone in the meandering Thames. Although the road does deviate from this direct alignment to cross the Medway and again to avoid the marshy ground along the Darent, it is clear that a direct route was first set out between Canterbury and Thorney Island.

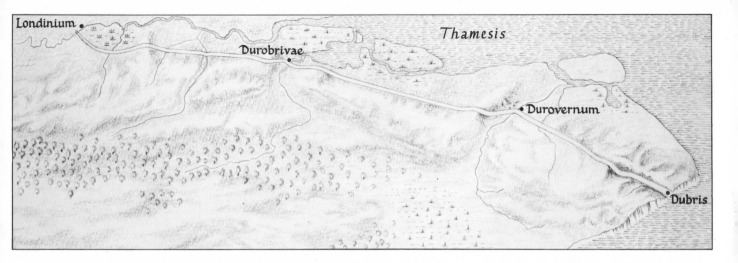

Londinium · · Durobrivae · Durovernum · Dubris · Thamesis

Just how the surveyors managed this is a mystery. It is relatively easy to connect two places which are visible from each other across flat, open country even without a map; but it is much more difficult to set out a straight line in wooded or hilly country between two points hidden from each other, or over long distances in any type of terrain.

Although we have no reliable evidence, it seems inevitable that setting out the road involved two separate stages. The first task was to establish the route. It is very likely that a line of beacons was used, perhaps by night, but more probably at dawn or dusk. From any beacon it would be possible to see the next in each direction and by some process of laborious adjustment they would be moved into a straight line to form a primary alignment.

The next task would be to transform this ideal line into a practical route on the ground. Where the terrain between two primary stations—beacons, in our account—presented no major obstacles, it was a simple matter of setting out the primary alignment with markers, such as stakes or stones at close intervals.

Where, by contrast, a wide river or unusually difficult ground lay between the primary stations, the line might be varied to find an easier route. So Watling Street makes a wide detour to cross the Medway at Bridge Reach; similarly on the section between Dover and Lydden Hill the road has first to run up the east side of the River Dour to avoid the numerous

Watling Street was cut through unmapped country of forests, rivers and marshes; here the artist has tried to suggest what it was like.

spurs and brooks on the west, and then to cross over just north of Temple Ewell as the east bank becomes too steep. As it happens, these particular problems of terrain do not cause the road to deviate very much from a primary alignment between Dover and Canterbury, but at Lydden Hill itself a considerable departure from this straight line is necessary to ascend the hill by the only practical route. From the top of the hill the Roman engineers evidently did not think it was worth returning to the original line and so struck a new straight alignment direct to Canterbury.

Beacons would naturally be set up on high points, so it is here that we often find very slight changes in direction. Thus between Canterbury and Rochester, Watling Street runs straight from the high ground at Harbledown Lodge three kilometres (nearly 2 miles) to another hilltop at Dunkirk, then turns very slightly and runs straight again to Chatham Hill, thirty kilometres (19 miles) away, with only trifling adjustments on the hillocks at Norton Ash and Radfield. Parish boundaries have frequently followed Roman roads and can form a valuable guide to the location of a vanished section of road.

21

left: *The groma. A surveyor would plant it firmly in the ground, check by the plumb-lines that it was perfectly upright, and sight along the arms or cords to mark out a straight line or right-angle.*

plumb-line

right: *Vitruvius' specification for a good road required a total depth of more than a metre.*

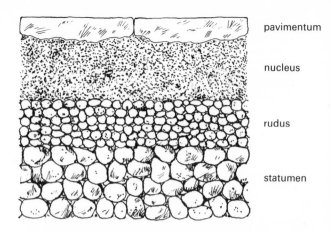

pavimentum

nucleus

rudus

statumen

The structure of the Roman road

The earliest Roman roads, which were rather like walls built on their sides, were made of stone blocks massive enough to stay in position simply by their own weight. Later roads were very varied, although occasionally similar stone causeways were constructed in different parts of the Roman world.

Vitruvius, who wrote on roads as on every other kind of engineering, has left us an account of the ideal road: according to him, it should have four layers, the *statumen*, *rudus*, *nucleus*, and *pavimentum*. Here, as in other matters, the genius of the Romans lay in constantly adapting themselves to their changing needs and local resources.

The strength of a road lay in its foundation, the statumen. Every sub-soil required a different foundation: the hard soils of North Africa needed very little, the rocks of the Alpine passes none at all; in the softer soils of much of Europe a solid foundation was essential to prevent the weight of traffic from destroying the road. It was usually sufficient to lay a course of broken stone, rammed in layers, but in marshy ground this would have had to be contained between two rows of wooden piles. In the worst bog, the whole road might be supported on a raft of logs and brushwood. Many such roads were built in the Low Countries.

In 1962 the demolition of three houses in Rochester High Street gave the Kent Archaeological Society an opportunity to study the structure of Watling Street. Their excavation revealed several roads, each laid on top of an earlier one which had been broken up by traffic or had sunk into the soft ground.

Some two and a half metres (8 feet 3 inches) below the modern road they discovered a layer of large flints, overlaid by a layer of sandy clay and another of rammed chalk. The depth of these three layers was about sixty centimetres (2 feet). The chalk may have been the surface of the earliest road, but

this is doubtful since it was still straight and level: the layer of gravel and cobbles which lies above the chalk is more likely to have formed the surface, since the cobbles were very uneven and awry as if worn by traffic.

At some time another road of gravel with a pebble surface had been laid on top of the dislodged cobbles, followed in its turn by a third road of rammed chalk with a cobbled surface, and finally a fourth road surface of gravel – a metre and a half (5 feet) of Roman road in all. This process has continued to the present day; for two thousand years road-stone has been carried to keep Watling Street fit for the tremendous traffic between the Channel ports and London.

Except where a road is carried on special foundations through water-logged ground, it is vital that rainwater be led away from it as quickly as possible; for both surface and foundations will be broken up if water is allowed to stand on the road and then to sink through into the soil below. All Roman roads were therefore cambered or sloped, so that rainwater would quickly drain away from the surface. A gentle slope on either side of the road then led the water to a ditch (*fossa*), usually two or three metres (7–10 feet) away, over ground which had been cleared of vegetation. It so happened that the very first trunk road which the Romans needed, the Via Appia, lay through the ill-famed and unhealthy Pontine Marshes. The fossae here must have resembled the drainage dykes of fen country. The Romans liked to raise their roadway on a bank, or *agger*, often no more than a metre (3 feet) high but occasionally much higher, both to aid drainage and to give marching troops a good view of their surroundings.

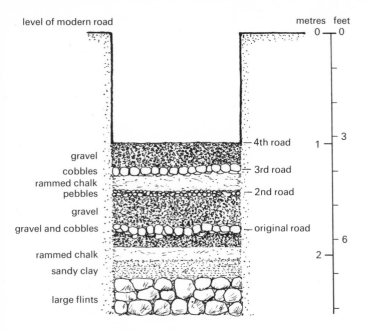

level of modern road

metres feet

0 — 0

gravel — 4th road
cobbles — 3rd road
rammed chalk
pebbles — 2nd road

1 — 3

gravel
gravel and cobbles — original road

6

rammed chalk

2

sandy clay

large flints

High-quality surfaces

The pavimentum, or *summa crusta* as it was sometimes called, had to be hard and smooth; hardness depends on the quality of the stone used, smoothness on the roadmakers' skill. On a few roads, of which the Appian Way is one, the surface was of slabs with a smooth finish carefully bedded in a nucleus of sand or sand and lime. These thick slabs were not like our paving-stones, but, like the facing stones in a Roman wall, tapered downwards to lock more firmly into the nucleus. Much more frequently, the wearing surface would be of gravel beaten down or rolled with huge stone or wooden rollers drawn by men or animals to form a dense, smooth surface.

The camber of the road acted as a first defence against water, and a well-paved surface left no cracks for rain to penetrate, besides being smooth for traffic. Where the slabs were rectangular, as on the limestone road between Antioch and Aleppo, it was relatively easy to get a tight fit; on the other hand fitting together the polygonal stones of the Via Appia must have resembled a giant jig-saw puzzle, with each piece as heavy as two men could lift. The surface of a gravel road needed a suitable mixture of fine and coarse material which would bind together when compressed.

left: *The section of Watling Street excavated at Rochester.*

below: *Different Roman methods of fitting together the stones of a road surface can be seen clearly in these two photographs from North Africa. One shows a road passing a milestone in the country, the other the road in front of the triumphal arch in the city of Timgad.*

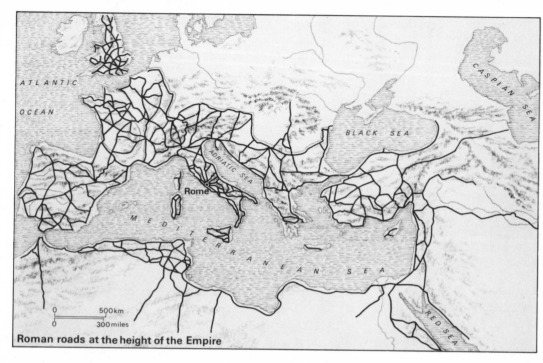

ATLANTIC
OCEAN

CASPIAN SEA

BLACK SEA

ADRIATIC SEA

Rome

MEDITERRANEAN SEA

RED SEA

0   500km
0   300 miles

Roman roads at the height of the Empire

below: A Roman milestone which stood by the Appian way, 13 miles (19.2 km) from Rome.

Arteries of the Roman world

We must always remember that Roman roads were built for specific purposes, to meet particular needs. The men who planned these roads, whether soldiers, politicians or, most probably, both, would not see swamps, deserts or mountain passes as challenges to their technical skill, as a modern engineer might, but simply as inevitable obstacles to an overall plan. The roads were the military and civilian chains linking the whole Empire together. These great highways started at the gates of Rome; in the city itself the streets were narrow, twisting and often ill-cared for. Many second-class roads in the provinces would also be poor. Thus the main reason for building trunk roads with smooth, hard surfaces, easy gradients and good visibility was to enable an army, composed largely of foot soldiers carrying everything on their backs, to cover great distances at speed.

Officials, inspectors and messengers on horseback or in fast carriages used the roads, and government rest-houses were placed along many routes. All kinds of non-official traffic also used the roads, from local farmers with their carts to merchants

An artist's reconstruction showing road building. The fossae are being dug ahead of the road to drain the water from the land to be excavated; the surveyor works ahead, too, so that clearing of vegetation and rocks can proceed along the correctly aligned route.

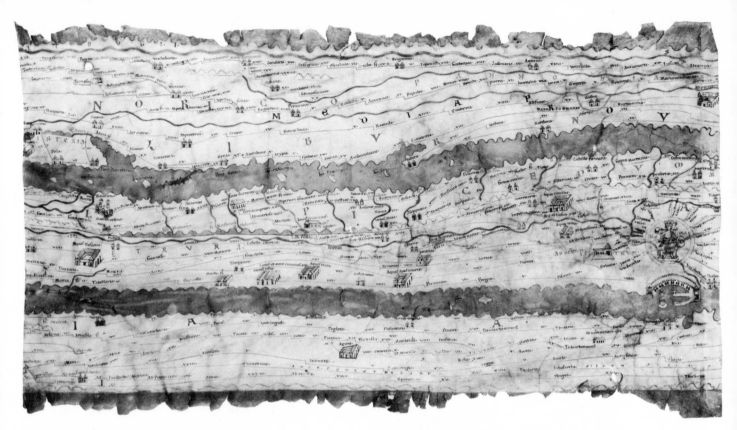

This is a mediaeval copy of a Roman road map. The central strip of land represents Italy, from Rome in the south to Ravenna in the north. The strips of water above and below represent the Adriatic Sea and the Mediterranean respectively. The upper coastline is that of modern Yugoslavia, and Trieste would be at the extreme left; the lower is the North African coastline with Tunis at the extreme right. Buildings indicate special features of certain places whilst camps or small towns are shown by deviations in the line of a road. Mileages are given and rivers and mountains are indicated.

The enlarged section shows Rome, symbolised by the Emperor, and Claudius' harbour at Ostia (see page 28).

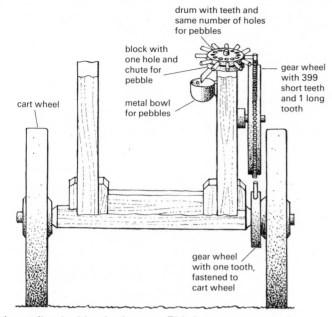

drum with teeth and same number of holes for pebbles

block with one hole and chute for pebble

gear wheel with 399 short teeth and 1 long tooth

metal bowl for pebbles

cart wheel

gear wheel with one tooth, fastened to cart wheel

A cart fitted with a hodometer. This instrument caused a rounded pebble to fall into a metal bowl after each mile. The cart had special wheels, four Roman feet (1.2 m) in diameter, twelve and a half Roman feet (3.7 m) in circumference. 400 revolutions made a Roman mile. The drawing is a reconstruction from a description by Vitruvius.

bringing wares from other provinces.

No one name is particularly associated with road-building, as Sextus Julius Frontinus is with aqueducts. The biographer Plutarch, however, writing of Gaius Gracchus, a politician of the second century BC, tells us that he introduced legislation for road-building and himself supervised construction. He tells us, too, that Gracchus had every road measured in miles and marked with milestones.

Apart from milestones, later travellers had the assistance of route maps. These were not true-to-scale topographical maps, but rather schematic plans of particular roads, detailing routes and mileages and showing towns and inns, mountains and river crossings. The Peutinger Table shown opposite is a fifteenth-century version of such a map.

Let us end with a Roman poet's view of road-making. Here is how Statius described it in the first century AD:

below: *Most traffic on Roman roads naturally went at walking pace, but a few references have survived from Roman writers that show us the time it took to get official messages from one place to another. On good roads, with frequent changes of horses, the messengers could keep up high speeds over long distances, but sea crossings could be fast or slow according to the weather.*

Some Roman Journeys

Date	Details and time of journey
AD 4	Special messenger from Lycia to Rome. About 3,100 km (1,930 miles) in 36 days.
AD 31	Imperial courier from Rome to Antioch by sea in bad weather. 2,500 km (1,550 miles) in 3 months.
AD 43	Emperor Claudius, *en route* for Britain, from Marseilles to Boulogne. 870 km (540 miles) in 10 days.
AD 68	Special messenger from Rome to Clunia (Central Spain). About 2,000 km (1,240 miles) in $6\frac{1}{2}$ days.
AD 68	Imperial courier from Rome to Alexandria by sea. 2,000 km (1,240 miles) in 28 days or less.
AD 69	Special messenger from Mainz to Rheims to Rome. Over 2,100 km (1,300 miles) in about 9 days.
AD 193	Imperial courier from Rome to Alexandria by land. About 3,500 km (2,180 miles) in 63–4 days.
AD 238	Imperial courier from Aquileia (by modern Trieste) to Rome. 750 km (470 miles) in 3–4 days.

'Here first a furrow plough to mark the verge,
Then deeply trench and excavate the soil;
Refilled, this ditch becomes foundation for
The spine above, lest worthless soil collapse
Or hard-pressed stone the fickle ground betray.
With rammers here, and here with close-set pegs
Now tightly clench the way. How many work
United! Here divesting naked hills,
Destroying thickets; splitting boulders there,
Or felling trees with iron. Workers here
The dismal tufa set in tempered soil,
While others drain the thirsty ponds and streams.'

left: *A relief carved about AD 200 showing ships in the port of Ostia. The sculptor has shown the divine powers that can bring good or bad fortune to the sailor—Neptune is in the centre—but he has also shown practical details very clearly. The ships are roomy cargo-carriers of the type that brought grain from Alexandria, and we can even see how their riggings was worked. Behind the ship under full sail, beyond the quay-side buildings, we can see the beacon at the top of the lighthouse. The other ship, sails furled, is being unloaded.*

below: *The port of Ostia. There were mooring facilities for more than a hundred ships in Trajan's basin. Goods were stored in the surrounding warehouses. The lighthouse probably stood on the island between the two moles.*

Ports

Many roads ended at the sea. The Mediterranean occupied a central position in the Roman world; it could be an obstacle or it could be a broad highway. As the Romans spread their power round all the shores and swept the sea free of pirates, trade thrived, ports had to be enlarged and new ones built.

Some of the most impressive engineering works were at Rome's own seaport, Ostia, at the mouth of the Tiber. As the city's population increased, so did the need for supplies, especially grain from Egypt. In the middle of the first century AD, the Emperor Claudius created a new harbour on the coast to the north, linked to the Tiber by a canal. It was formed by two curved moles jutting into the sea, with an artificial island between their ends to create two entries to the harbour. This enclosed about 70 hectares (0.27 square miles) of sea, but it proved to be too exposed in bad weather. So, half a century later, Trajan had a basin dug inland, about half the size of Claudius' harbour.

Few ports could compare with Ostia, but very many required constant skill and labour to improve and maintain them: quays, piers and moles, warehouses and lighthouses were needed everywhere. It is sometimes forgotten that the Roman seaways were as important as the Roman roads, and that they equally depended on good engineering.

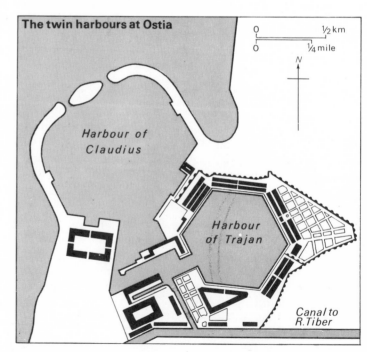

The twin harbours at Ostia

Harbour of Claudius

Harbour of Trajan

Canal to R.Tiber

4 Bridges

The army's wooden bridges

Bridges are important features of both aqueducts and roads. Although the Romans were not the first bridge-builders, any more than they were the first to build aqueducts or all-weather roads, their work was unique in quality and scale. The Romans' outstanding achievement was the arch bridge, which they brought to perfection.

In the sixth century BC they had crossed the Tiber with the first Pons Sublicius, the simple wooden bridge which Horatius is supposed to have held against the Etruscans. Even later, when stone arch bridges were common, they continued to build timber bridges, such as Caesar's famous military bridge across the Rhine, described in his *Gallic Wars*.

'For these reasons Caesar had determined to cross the Rhine, but decided that to cross in boats was both too risky and beneath the dignity of the Roman people. So although it would be extremely difficult to build a bridge, since the river was broad, swift and deep, he decided to make the effort or not to take his army across at all.

'He designed his bridge as follows. He fastened together a pair of piles, one and a half feet (45 cm) thick, pointed at the lower end to suit the varying depth of the river, spaced two feet (60 cm) apart. These he lowered into the river with tackle, made firm, and drove home with pile-drivers, not vertically, as piles usually are, but sloping obliquely, inclined in the direction of the current. Opposite these, at a distance of forty feet (12 m) from the lower part, he set up a further pair, similarly fastened together and inclined against the flow and current of the river.

'The distance between each pair was maintained by a beam, two feet (60 cm) thick, placed above and fitting between the two piles, secured by a pair of braces at each end . . . Once all these trestles were in position, planks

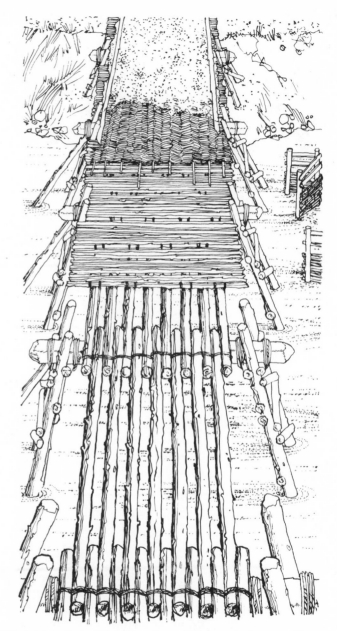

This bridge was built by Caesar's army to cross and recross the Rhine; then it was destroyed. The drawing has been adapted from an imaginative drawing of the bridge by the sixteenth-century Italian architect Palladio, based on Caesar's own description.

The Column of Marcus Aurelius (AD 121–80) in Rome, like that of Trajan, has a spiral relief showing his achievements. Here soldiers are crossing a pontoon bridge or a 'bridge of boats'.

were laid between them, surmounted by long poles, and covered with hurdles.

'Other stakes, moreover, were placed at an angle downstream and secured to the rest of the structure to act as props and to take the force of the current; others, too, were placed a little upstream of the bridge, so that if the natives tried to demolish it by sending down tree-trunks or boats, these buffers would absorb the force of the impact and prevent any harm to the bridge.

'Ten days after the collection of the timber had begun the bridge was finished and the army crossed over.'

Many smaller bridges of this sort were built by the army, who were skilled, too, at constructing pontoon or floating bridges.

Stable stone arches

Let us now look closely at the arch. Before the Romans, the Etruscans had built crude but effective arches in Italy; the Romans were to develop the skill to an extremely high standard. We can reconstruct in drawings of four simple experiments the experience of the Romans over several hundred years. An essential feature of any arch is the temporary wooden structure, known as centring, which has to remain in position supporting the arch until it is finished.

These experiments reveal three of the four essentials of a Roman arch: stout abutments or piers at the sides, properly shaped centring, and well-fitting voussoirs. (The voussoirs are the wedge-shaped stones that make up the arch.) Obviously the arch also needed a solid foundation; without this it would push over even strong piers if they were built on soft ground. Much thought had to be given to the siting of foundations, since many Roman bridges carried a road over a river, and river banks are difficult ground for building.

Establishing foundations

At the beginning of a road-building project the censor might have to alter his ideal line to meet the engineer's needs. A detour could be justified if it allowed a narrower crossing or produced firmer ground. The engineer would direct his surveyor to measure the width of the river at several points, and his labourers to dig trial-holes to see what the sub-soil was like.

We cannot say for certain how the librator would measure a wide river too deep to be forded, but he may have thrown a rope across or he could have calculated the width from the bank using Euclid's geometry, with which Roman surveyors would be familiar.

Once a site has been chosen and the abutments marked, work can begin on the foundations. The chief problem is, of course, water; as soon as workers begin digging for an abutment, water will seep into the hole, and in the case of piers there are even more difficulties because their foundations have to be laid under the river-bed itself. However, both problems can be solved with a coffer-dam. A coffer-dam is a watertight box built of piles, a little bigger than the projected foundations. The bottom is driven firmly into the ground; the top is open. It is built around the place to be excavated and the water caught inside the dam is pumped out.

The Romans would build their coffer-dams from timber; there would nearly always be double walls, as in this one

Making a stable semi-circular arch

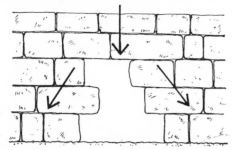

The weight of a wall thrusts downwards. It is possible to make a hole by removing some blocks so that the others still support one another and carry the thrust outwards and downwards to the ground.

The hole can be made neater, and wedge-shaped blocks (voussoirs) fitted to hold one another together at the top of the hole. This makes a rough semi-circular arch. While the voussoirs are being fitted, they need to be supported by a wooden frame, which is removed when they are all holding one another together.

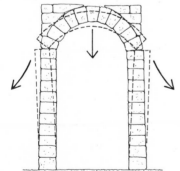

If the voussoirs are carefully shaped, the arch will be stronger and the opening can be made bigger. But if the sides are too thin and light, they will be thrust out, and the whole structure will fall down.

So it is vital that every arch has plenty of side support, either from a wall or massive piers.

31

This bridge, known today as the Ponte Grosso, formed part of the repairs to the Via Flaminia ordered by Augustus around 27 BC. It spans a river within the Furlo Gorge near Urbino and its abutments and piers were made very strong to withstand storm waters.

described by Vitruvius. 'Double coffer-dams, bound together with planks and chains, are to be put in, and clay is to be pressed down as tightly as possible between them. Then the enclosure is to be emptied of water with waterscrews and waterwheels.'

Vitruvius must be thinking of a big foundation in deep water, since waterscrews and waterwheels would be large and expensive. In most cases buckets or a bucket-chain would be adequate. The waterscrew, or Archimedean screw, however, was particularly useful, as it is today, since it could be turned by slaves working on dry land or in a barge tied up alongside the coffer-dam.

Even once the area inside the coffer-dam is dry, the engineer's problems are not over. If he finds a solid base, he can simply fill the enclosure with concrete; if not, he must, says Vitruvius, provide such a base: 'If a solid foundation is not found, piles of olive, alder, or charred oak are to be driven

close together by machines and the space between them filled with charcoal.'

By 'charcoal' Vitruvius means wood which, like the oaken piles, has been charred to prevent rotting. This would be packed between the pile-heads to make a solid timber platform on which to start building the foundations of the piers.

The earliest stone bridges were built of massive stone blocks, accurately dressed and joined without mortar. Such blocks would be very awkward to handle from a barge, so wherever possible a river would be bridged by a single span at a narrow point to avoid the difficulty of building piers in mid-stream. In time, however, this problem became less important, since the invention of mortar caused a gradual change in building methods. The strength of finished masonry came to depend less on the weight of the individual blocks than on the bonding-power of the mortar, which allowed the use of much smaller stones.

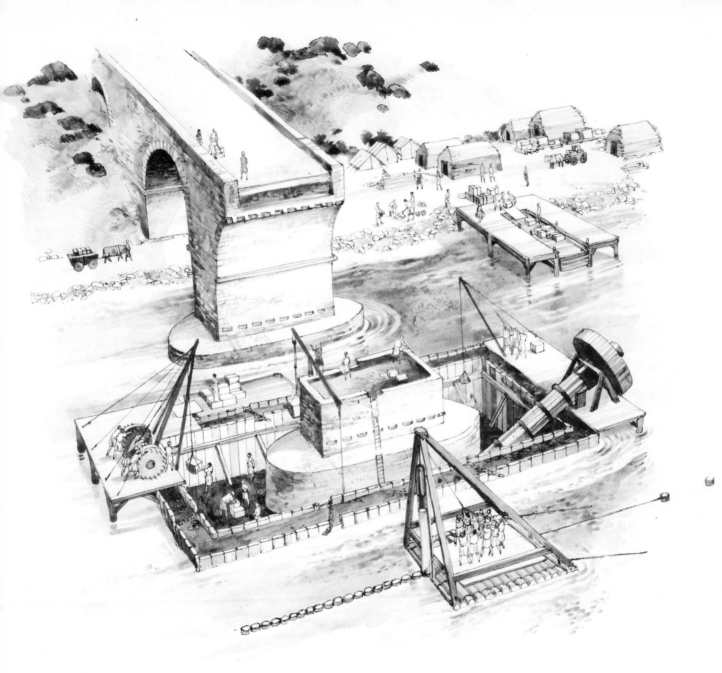

Building a bridge: an artist's reconstruction to show the different tasks and machines involved. The cofferdam of timber piles with a water-seal of clay has enabled the builders to carry *the second pier of the bridge well above water level; soon it can be dismantled and the work can continue from scaffolding built around the pier.*

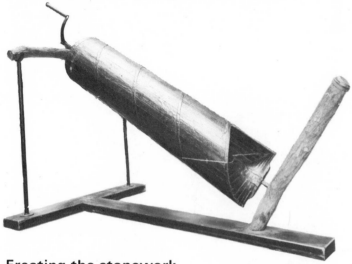

below: *Archimedean screw. When the handle is turned, the long wooden spiral inside the tube pulls up water from the lower level to the upper. This one is about average size for Egyptian irrigation work, 2 metres (80 inches) long with a diameter of 46 cm (18 inches). It was made and used in Egypt about 1920, but the design has hardly changed since ancient times.*

above: *This Augustan bridge was built to cross the swift-flowing River Nera at Narni. Although now unsupported on the right, the arch is still stable because of its substantial pier. You can see clearly the projecting corbel stones on the lower half of the pier.*

Erecting the stonework

Look carefully at the bridges on these pages; several features give clues to the method of construction. Projections of various kinds can be seen on all of them, either isolated stones or complete courses. This corbelling had two main functions: to support the scaffolding and to carry the centring. The single projecting stones on the bridge over the River Nera at Narni in Italy were provided for the scaffolding, which may have been completely supported by the corbel stones, or built up from the ground and simply anchored to them. The corbels were rarely cut off afterwards—after all, they might be useful for inspection and repair of the structure.

The masons would build the piers to just above the springing level. At this point the carpenters would take over and begin to construct the centring—the great timber framework, carried on the highest corbels, which had to support the voussoirs until the last one was firmly in position. The semi-circle of boards secured to the outside of the framework had to be accurately shaped to the soffite (underside) of the arch. Each timber component of the centring had to be exactly cut and secured in place by iron spikes, for many tonnes of stone would

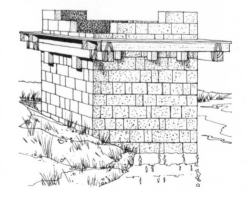

left: *Scaffolding could be built up from stones which were left projecting so that there was no need to carry vertical supports down to the ground.*

34

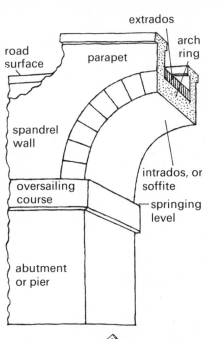

road surface

parapet

extrados

arch ring

spandrel wall

oversailing course

abutment or pier

intrados, or soffite

springing level

right: *The Pons Aemilius, Rome. The massive foundations of this bridge protected it against the Tiber's floods for 1,600 years until 1575 and they still protect the one remaining arch.*

left: *Engineers have special names for the parts of an arch.*

below: *The timber framework, known as centring, takes the weight of an arch until the last stone is inserted. Then the wedges are eased out to allow the framework to drop and the stonework to take its own weight.*

completed, to transfer the great weight of the stone into the arch itself by easing the wedges thus allowing the centring to drop away from the stonework.

Raising the heavy voussoirs to that height would not have been easy. Many types of hoist were used, but we know of one in some detail from a funerary relief (page 36). You will see that power comes from slaves turning a treadmill; mules were unsuitable, since stopping and starting had to be controlled exactly.

When the final voussoir is in position, the anxious moment comes when the order is given to ease the centring. Will the arch stand or will it twist itself out of shape? The answer depends entirely on the accuracy of the masons' work. The wedges are out, the centring drops a little and the great arch settles gently into the graceful curve it will hold for many centuries to come! Even the slaves cheer!

The centring will not be completely removed until all the work is finished. Spandrel walls must be built and, between them, the curve of the extrados smoothed out with rubble or concrete. Then the arch will be complete and workmen can mount it to finish the bridge: build the parapets, add copings and ornaments, and lay the road surface.

have to be supported. Most probably the framework would be built first at ground level, to ensure accuracy, and then dismantled and re-erected on the stonework.

Once in position, the centring would not rest directly on the corbels, but on several pairs of folding wedges. These would allow the engineer, once the curve of the arch had been

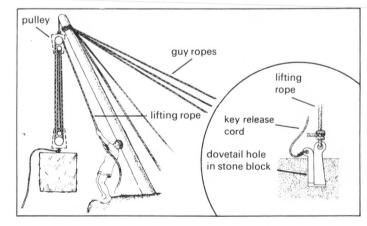

The finest bridge in the Roman world (shown on the cover) crosses the River Tagus at Alcántara in the Estremadura region of Spain. It is built entirely without mortar. The six mighty spans rise nearly fifty metres (165 feet) above the river and the bridge itself is almost two hundred metres (655 feet) long.

It is strange to recall that the forebears of the men who built at Alcántara, Segovia, and at Nîmes were so fearful of the spirits which they believed haunted every corner of their native Latium that their *pontifices* had to placate these spirits with ritual before the humblest stream could be crossed with a log; the Latin *pontifex*, priest, originally meant 'builder of bridges'.

far left: *A relief dated about AD 100, found on a tomb near the Porta Maggiore, Rome, showing the construction of a temple. A crane is being erected. The men at the top are lashing the guy-ropes to the jib while others are climbing into the tread-mill ready to work it. The two-sheave pulley, one wheel above the other, corresponds exactly to a description given by Vitruvius. This type of crane would have been used to lift heavy stones in many large Roman engineering projects.*

left: *The artist's drawing shows how such a crane could be used to raise large stones to a considerable height. The workmen in the picture are completing the spandrel wall of a bridge.*

below left: *A lewis is a device for attaching a crane rope to a stone block. The mason cuts a dovetail hole in the top of the block; a wedge-shaped metal anchor is placed in the hole and locked there by a parallel metal key. When the block is in position the key is pulled out; by attaching a cord to the key this can be done under water.*

right: *The Spanish town of Segovia is dominated by this magnificent two-tier aqueduct 728 metres (2,388 feet) long and 29 metres (95 feet) high above the road. It was completed early in the second century AD and is still in working order.*

How the Romans adapted the arch

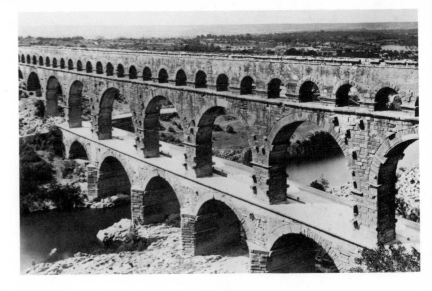

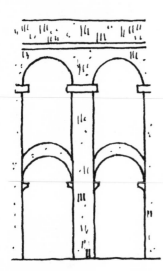

Placed side by side, arches could form an arcade of almost any length, to carry a road or a water channel. If extra height was needed, one arcade could be built on top of another; an example of this is the Pont du Gard, near Nîmes in southern France, which takes an aqueduct across a deep valley.

Another way of achieving extra height was to insert supporting arches to prevent very tall pillars from bending sideways; the aqueduct at Mérida, in south-western Spain, is strengthened in this way.

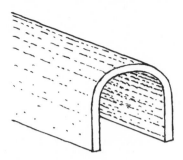

Imagine arches now placed, not side by side, but one behind another to form one very deep arch; this is a plain vault, and there is an example in the cryptoporticus which the Emperor Nero built in his Golden House in Rome, about the middle of the first century AD.

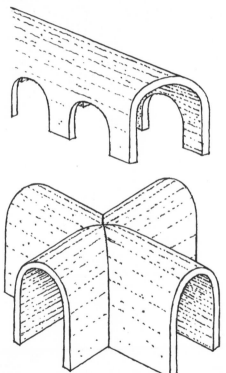

Vaults could have arches cut in their sides, and this idea could be developed so that one vault cut right across another of the same size; the vaulted passages inside the Colosseum show how one side could be a row of arches, yet still strong enough to support a huge weight.

The Colosseum as a whole—or the Flavian Amphitheatre, to use
its correct name, because it was built by the Flavian emperors
late in the first century AD—is an enormous pattern of arches,
arcades and vaults fitting together in a structure strong enough
to hold 87,000 spectators and everything that was needed for
the shows.

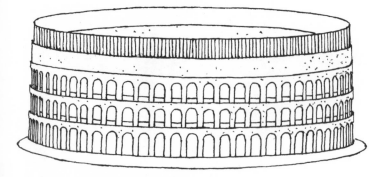

If we imagine a series of
equal-sized arches crossing
one another at the centre, we
have a dome.

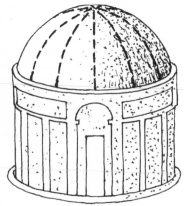

right: *Even when the Roman Empire in the West had collapsed,
its engineers' skills lived on. The church of San Vitale, Ravenna,
was built about 530 and has a central dome supported all
round by arches and walls pierced by arched windows; the
whole complex arrangement can be seen in this outside view,
though the dome itself is hidden under a tiled roof.*

5 Materials

Stone

It is clear that stone was the most important Roman building-material, and indeed it is quite remarkable how much was done with nothing but dry stone. Foundations were laid, water-channels, walls, bridges, and vaults built, and roads surfaced. All this reflects the great skill and care that the Romans applied to quarrying and shaping. In one way, their eventual progress to mortar jointing was a step backwards for the masons' art; they no longer needed to cut and shape so carefully.

The Romans were fortunate in their local stones. It was almost as if Nature had set up a school for masons in the country around Rome. In the very early days the nearest stone was the soft *cappellacio*, formed from the dust of ancient volcanoes, which could be quarried with ease and shaped with inferior tools. But it was not durable and soon crumbled under the effects of weather. As Rome overcame her neighbours, she was able to draw on harder, more durable volcanic stone from the Alban Hills and lower Appenines, using masonry techniques already perfected on the softer stone. The Romans also learnt that stone weathers better if it is laid in a building the same way up as it was in the quarry.

The masons developed other tricks, too. They could, for instance, fit the blocks together more tightly along the joints on the face of a wall if they slightly tapered the sides towards the back. An even more cunning device was called *anathyrosis*: the end faces of each block were slightly hollowed so that all four edges were sure to touch. The four edges of the front face were often chiselled smooth to allow the mason to place his measuring square at each corner to check it had been accurately cut. The rest of the front face could be left rough and unchiselled since it was not going to be laid flush against another stone. Although all these devices improved the appearance of masonry, their main function was to strengthen it and exclude water.

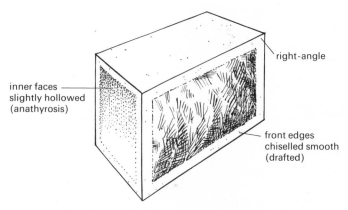

above: *A carefully prepared stone building block. It would have been about half a metre (one and a half feet) square by one metre (three feet) long.*

below: *A typical composite wall with squared stone faces and a rubble core (see page 42). On this wall the rough outer faces of the stones have been chiselled smooth after being fitted into place.*

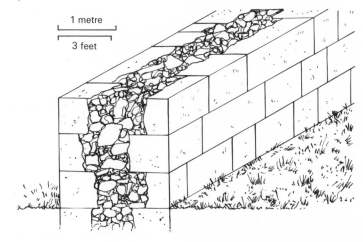

From the earliest days the blocks of stone were laid in varying lengths and widths with their vertical joints staggered at random. A wall would be weaker if the vertical joints were directly above one another. Only the depth was kept the same. The variety of width of the stones did not matter, since a wall was usually built with two skins of stone, each with a flat outer face, and the cavity between was filled with rough rubble.

Mortar began to be used in joints during the second century BC. Early mortar was weak, and spread thinly between accurately dressed blocks. However, when the Romans had discovered how to make stronger mortar, stones no longer needed to be so carefully cut. This new mortar, which had greater bonding power, gave a number of smaller stones the strength of a single large block.

Mortar and concrete

The magic material which has given the Romans an everlasting reputation for strong mortar is called *pozzolana*. The Romans thought it was a special kind of sand, but it is in fact ash from prehistoric volcanic eruptions. Mortar was strongest when made from pozzolana deposited close to the volcano. Wind-borne ash which had been carried great distances tended to become mixed with soil; this could affect the strength of the finished mortar. Pozzolana takes its name from the Italian town of Pozzuoli, known in Roman times as Puteoli, from which came its Roman name, *pulvis Puteolanus*.

The Romans took a long time to discover that pozzolana could be found at many places much nearer to Rome than Puteoli. They had been deceived by its variety of colour–from red to brown, grey, or almost black–which depended on the particular volcanic activity which originally produced it. Roman mortar used pozzolana just as ours uses sand; two or three parts of pozzolana were added to one part of lime.

Builder's lime is made from limestone burnt in a kiln; there was plenty of limestone near Rome, but like pozzolana it varied in quality from place to place. The experience gained by generations of Roman masons in combining these two variable materials into a first-class mortar helped later generations to build to the same standard when working with unfamiliar materials far from Rome.

Around 1910 a wall-painting was discovered in the tomb of Trebius Justus on the Via Latina which shows a workman mixing mortar in a wooden trough known as a *mortarium* (from which comes the word 'mortar') with a broad-bladed hoe called a *rutrum*. The workman is said to have 'an anxious expression as though well aware that the strength of the wall depended largely on the excellence of his mortar'. This would certainly have been true if his mortar was intended for concrete rather than for jointing stones or bricks. For once mortar had been discovered, it also became possible to make concrete. Roman concrete was not suddenly invented; it developed from the practice of building walls with a rubble core. The addition of a good quality mortar to the rubble filling made the wall much stronger; in due course, this process became systematic and carefully controlled so that the same high quality could be

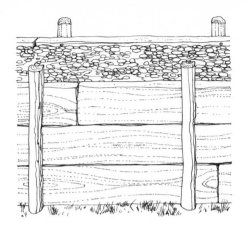

left: *A concrete wall being built between wooden shuttering; this is being left in place until the concrete sets.*

below: *A wall in the Teatro Piccolo (small theatre) at Pompeii showing both brick and stone facing. The stonework under the arch is opus reticulatum but weathering has taken the sharp edges from the stones.*

reproduced every time.

One method was almost universal: a layer of small stones, between thirty and sixty centimetres (one and two feet) deep, was followed by a layer of mortar, which was lightly rammed down to flow between the stones; more stones were laid on top, followed by another layer of mortar, and so on, making a wall of concrete. Where the concrete was being placed between brick or stone face-work, vigorous ramming would have pushed the face-work out of place. Some concrete walls were unfaced but built between temporary wooden faces, which we call shuttering. Marks left by these boards are occasionally found on Roman concrete.

The permanent facings of stone or brick are of great interest and variety. Where ordinary squared masonry was used, the wall appeared to be built of solid stone. More often, however, the stone facing was in the form of *opus incertum* or

opus reticulatum. (These two Latin names are not Roman but mediaeval.)

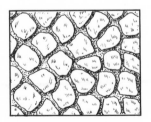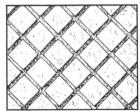

above left: *Opus incertum: irregular small blocks of stone.*

above right: *Opus reticulatum: regular small pyramid-shaped slabs of stone in a fish-net pattern.*

Bricks and tiles

Wherever there was suitable clay and the climate was sunny enough, early civilisations made crude bricks baked in the sun. Vitruvius said that a good sun-baked brick should have been exposed for two years. Roof-tiles, *tegulae*, are also made of clay, but since sun-baked clay is broken up by water it is not suitable for roofing. So roof-tiles were the first to be fired in a kiln, and it is here that the story of Roman bricks really begins.

This relief was found at Lake Nemi in the Alban Hills, south of Rome, and dates from the first century AD. It shows part of a Roman town. Within massive walls of well-cut stone a crowd of buildings is packed in narrow streets. Some of them seem to be two or three storeys high, and only in parts has the sculptor tried to indicate stonework.

Ancient Rome was a crowded city of narrow streets. Most people lived in tall tenements called *insulae* (islands). Because they were often too high for their walls to support, these tenement blocks frequently collapsed. Also since they were heated by braziers and built very close together, they were often destroyed in widespread fires. Added to this, the vast schemes of public building involved extensive clearances. All this destruction produced an enormous supply of rubble; that from the walls and floors was of little use, but the fired-clay roof-tiles, which the Romans had early perfected, were often

recovered intact or broken into two or three pieces.

If their flanges were broken off, these large tiles made extremely serviceable building-blocks. Thus the Roman brick was born. Even later, when they set out to make fired bricks, the makers kept roughly to the proportions of the original roof-tile.

Roman walls were rarely built entirely of brick. The bricks had two main functions: one was to face the concrete core, and the other was to form occasional bonding-courses, running the whole length and width of the wall.

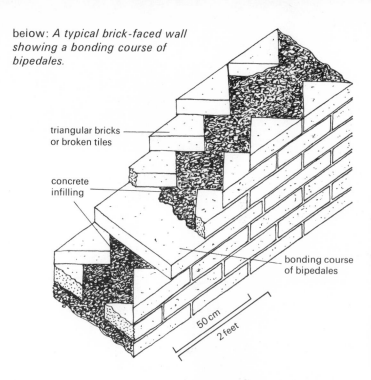

triangular bricks or broken tiles

concrete infilling

bonding course of bipedales

50 cm

2 feet

The drawing shows a typical Roman brick-faced wall, which uses two sizes of bricks. The shorter bricks are about forty centimetres (16 inches) long and five to nine centimetres ($2–3\frac{1}{2}$ inches) deep. They may have been cut from a roof-tile, or be purpose-made. In either case they will taper back into the wall, some to the extent of being triangular. In this way they are gripped more securely in the concrete core.

The larger bricks are called *bipedales*. They are sixty centimetres (2 feet) square and about seven centimetres ($2\frac{3}{4}$ inches) thick, and are as heavy as one man can lift unaided. Each bipedalis lies flat, one edge showing in each face of the wall, which is sixty centimetres (2 feet) thick. Each course of bipedales runs the whole length of the wall, and serves to bond it more tightly. Such courses, sometimes composed of several layers of smaller bricks, are also found in masonry walls.

The bipedalis often carries the name of the brickyard where it was made, and sometimes of the man who made that particular brick. Sometimes, too, and this is very useful, it is dated in the usual Roman way, by the name of the consuls (or local magistrates) for that year.

A Roman bonding brick from St Albans, measuring one by one and a half Roman feet (29.6 by 44.4 cm).

Metal and wood

Stone, concrete, bricks, and mortar were the Roman engineer's foremost materials, but metal was needed for tools, and wood for many temporary structures.

The Roman iron-industry was well developed throughout the Empire. By the time of Augustus iron was, according to Strabo, a contemporary Greek geographer, already an important export from Britain. Over the next century the industry expanded enormously, and iron was produced in greater quantities throughout Britain, especially in Sussex and the Forest of Dean.

Despite the scale of the industry, however, and the quality of the iron goods produced, the Romans never discovered that if iron ore is heated to the melting-point of iron, the metal will liquify and can be separated from the impurities present in the ore. When this happened accidentally, the liquid, which when

cool becomes cast iron, was thrown away as useless. The Roman iron-master removed the impurities from his iron by repeated hammering, so the objects he could produce were limited by the size of the lump of impure iron that a man could handle (usually about 25 kg, 56 lb). For this reason the use of iron in building was limited to such things as nails and cramps.

The Romans had no method for making steel. However, they discovered that wrought iron could be converted into a hard metal not unlike steel by increasing the amount of carbon it contained. This the Roman smith did by hammering charcoal into the iron when it was hot enough to glow cherry-red.

The steel produced could be given a good cutting-edge by grinding, and was used in every kind of cutting tool. Even so, it was, of course, not as tough as modern steel. Although

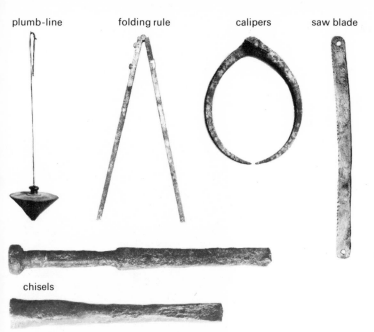

plumb-line   folding rule   calipers   saw blade

chisels

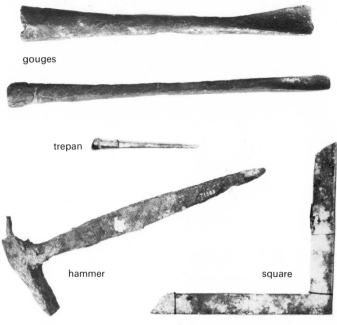

gouges

trepan

hammer   square

A model in Reading museum (opposite) shows a Roman carpenter in his workshop. The four main stages of his work are measuring, shaping, finishing, and joining.

His measuring instruments are shown here. The plumb-line enables him to set his work vertically. He has a folding rule and a set of external calipers for measuring awkward shapes.

For rough shaping he uses an adze, lying on the bench ready to hand. More accurate work will require a saw: the frame-saw hanging on the shelves will have a spare blade

since it is easily blunted. If holes are needed, the carpenter can use the auger hanging on the wall under the shelves or a trepan held vertically under the palm of one hand and turned with the aid of a bow in the other hand. He has a good array of chisels and gouges.

The carpenter is finishing a piece of wood with a plane, the wood held firm by bench-spikes. He will join it with nails or the glue he has ready in the pot, checking the joint with a square.

Roman carpenters possessed saws very like our own, it was extremely difficult to keep them straight and sharp, so they preferred to use an adze and plane wherever possible.

Despite these difficulties Roman carpenters played a major part in building-projects. Outside Rome most buildings, except the more important ones, were undoubtedly made of wood, but we know little of them since nothing has survived. Apart from wood for buildings, wood was of first importance on any building-site for all temporary works such as scaffolding, centring, or coffer-dams, and machinery such as cranes and waterscrews.

Vitruvius shows that great care was taken to select the timber appropriate to the task in hand. Alder, he tells us, though apparently useless in normal conditions, will last forever in water. The buildings of ancient Ravenna (which was built on an archipelago of lagoon islands rather like Venice) rest entirely on piles of alder.

Even Vitruvius, however, allows his imagination to run away with him sometimes. He regrets that larch, *larix*, which is found only on the banks of the Po and along the shore of the Adriatic, cannot be brought to Rome; for, he claims, it is entirely fireproof. The name, he says, comes from a town besieged by Julius Caesar called Larignum. The inhabitants of this town were defending themselves from a wooden tower so Caesar had faggots piled high around the base and set them alight. Though entirely obscured by flame, the tower stood unscathed. The astonished Caesar at once set himself to find out the name of the timber!

6   The Roman achievement

Aqueducts, roads, and bridges have been the foremost theme in this book because these above all demonstrate the Romans' grasp of the basic principles of engineering. They understood the need to make a thorough survey of the ground and a detailed study of a project before work began, and then to adapt their scheme to suit local conditions. They appreciated the importance of sound foundations and the equally important need, especially when building roads, to protect the foundations from water. In perfecting the arch, they realised that careful shaping and accurate cutting of the voussoirs would allow that semi-circle of stone to carry any load forever.

The application of these principles produced a vast amount of work that achieved a formidably high standard. Sometimes, however, construction did fall below standard. The ever-vigilant Pliny reports such a case to Trajan: 'the citizens of Nicomedia have spent staggering sums on an aqueduct, twice begun, twice demolished, but still they have no water-supply!' Ancient writers complain, too, of the dishonesty or carelessness of the builder, whose enthusiasm for tall insulae far outran his observance of structural principles, with disastrous results.

The Roman engineers' confidence in their knowledge and ability is magnificently expressed in such remarkable structures as the Colosseum (page 40). Here a most skilful combination of arch and vault has produced an arena in which many thousands of spectators could sit and—more important —move to and from their seats with ease. The boldly executed harbours of Ostia (page 28), the outer built under Claudius, the inner under Trajan, were inspired by this same confidence, as were the two long drainage tunnels driven through iron-hard rock, one from Lake Albano, the other from Lake Fúcino. But even the Romans could not predict everything accurately. Miscalculations did occur, as Tacitus tells: when the tunnel from Lake Fúcino was opened, the flow of water was so much greater than expected that it swept away the ceremonial banquet arranged to celebrate the opening!

This view in the amphitheatre at Pozzuoli shows both the engineers' highly developed use of arches in their design, and the strength of their concrete, standing firm after two thousand years even if some of the brick facing has fallen away.

Glossary of Engineering Terms Used in This Book

Note: Words shown in italic type are Latin.

abutment—a support at each end of a bridge

aedilis (plural *aediles*)—an elected official responsible for maintaining public works

agger—the bank on which a Roman road was built

anathyrosis—a slight hollowing on the ends of stone building blocks designed to create a tight fit between blocks

aqueduct—a conductor of water; a channel used to bring water to a city from a distant source

arch—a curved stone structure over an opening that supports the weight above it

Archimedean screw. See waterscrew.

bipedalis (plural *bipedales*)—a large brick used in Roman walls that extended from one side of the wall to the other

camber—a slight slope built into a road to allow water to drain from the surface

censor—an elected official who supervised building projects

centring (centering in American spelling)—a temporary wooded framework used to support an arch during its construction

chorobates—an engineering instrument in the form of a long, narrow table, used to check the alignment of an aqueduct channel

cippus (plural *cippi*)—a milestone used to measure distances along an aqueduct

coffer-dam—a temporary dam used to keep water out of a construction area. A Roman coffer-dam was a wooden box-like structure with double walls. The base of the dam was driven down into the bed of a river, and the water inside was pumped out by means of buckets or a waterscrew.

concrete—a building material made of small stones joined together with mortar or cement

corbel stone—a stone projecting from the surface of a bridge pier, used to support scaffolding or centring

curator (plural *curatores*)—an appointed official responsible for a specific category of public works such as water supply or highways

Curator Aquarum—Director of Water Supplies

Curator Viarum—Director of Highways

dioptra—a surveying instrument used to establish a horizontal line between two points

fabrus—(plural *fabri*)—a skilled craftsman in the Roman army who worked on engineering projects

fall—the downward slope of an aqueduct; the proportion between the length of the aqueduct and the difference in height from one end to the other

fossa (plural *fossae*)—a ditch on either side of a Roman road designed to carry away water

groma—a surveying instrument used to mark out right angles or straight horizontal lines

insula (plural *insulae*)—a Roman tenement or apartment building

librator (plural *libratores*)—a Roman surveyor

mole—a stone structure extending out from a coastline to form a dock or a breakwater

mortar—a building material used to join bricks or blocks of stone and to make concrete. Roman mortar was made of lime and volcanic ash.

nucleus—one of the two middle layers of a Roman road

opus incertum—small irregular stone blocks used in the facing of a wall

opus reticulatum—small regular pyramid-shaped stones arranged in a fish-net pattern on the surface of a wall

panniers—large baskets carried by pack animals. One of the mules in the picture on the back cover of this book is carrying two panniers.

pars inferior—the lower section of an aqueduct

pars superior—the upper section of an aqueduct

pavimentum—the top layer or surface of a Roman road, usually made of hard-packed gravel or flat stones

pier—a large column used as a support

plumb line—a string with a weight at one end used to check the vertical position of an object

pozzolana—a volcanic ash used by the Romans to make mortar

Acknowledgments

praefectus fabrum—a Roman army officer in charge of engineers and craftsmen

puteus—a vertical shaft from which work on a tunnel was begun

rudus—one of the two middle layers of a Roman road

specus—the water channel in an aqueduct

statumen—the foundation or bottom layer of a Roman road

tegula (plural *tegulae*)—a roof-tile made of fired clay

viaduct—a continuous row of arches carrying the above-ground portion of an aqueduct

Via Appia—the most famous of the Roman roads in Italy

voussoir—one of the wedge-shaped stones that make up an arch

waterscrew (Archimedean screw)—a cylinder with a spiral-shaped structure inside and a handle at the upper end. When the handle is turned, water is drawn from the lower end of the cylinder to the upper end.

Watling Street—the most important road in Roman Britain. It ran from Dover on the east coast to London.

Illustrations in this book are reproduced by kind permission of the following:
pp. 1, 7, 10 (above), 28, 30, 36, 40 (below), 44 Mansell Collection; pp. 4, 47 (plumb-line, calipers, folding rule, trepan, set square) Soprintendenza Archeologica, Naples; p. 5 drawn from information supplied by Chichester Excavations Committee; p. 5 Aerofilms; pp. 9, 10 (below), 18 (right), 24, 32, 39 (below), 42, 43, 45 (above), 48 Fototeca Unione, Rome; pp. 13, 35, 40 (above) Edwin Smith; p. 18 (left) Bath City Council; p. 19 J. Allen Cash Ltd; p. 23 (above) Gad Borel-Boissonnas; p. 23 (below) Michael Holford Photographs; p. 26 National Library, Austria; p. 34 (left) Crown Copyright. Science Museum, London; pp. 34 (right), 39 (above) Alinari; pp. 37, 38 (below) back cover Ampliaciones y reproducciones MAS; p. 38 (above) La Caisse Nationale des Monuments Historiques et des Sites; p. 45 (below) Verulamium Museum, St Albans; pp. 46, 47 (chisels, gouges) Reading Borough Council; p. 47 (saw blade) Swiss National Museum, Zürich.

Drawings by Peter Dennis and David Bryant
Maps by Reg Piggott

Measurements

Roman		English		Metric
1 *pes* (foot)	=	11⅝ in	=	29.59 cm
x 5				
=1 *passus* (pace)	=	4 ft 10¼ in	=	1.48 m
x 1000				
=*mille passus* (thousand paces, or a Roman mile)	=	1618 yd	=	1.48 km

Index

The Cambridge History Library

The Cambridge Introduction to History
Written by Trevor Cairns

The Cambridge Topic Books
General Editor Trevor Cairns

The Cambridge History Library will be expanded in the future to include additional volumes. Lerner Publications Company is pleased to participate in making this excellent series of books available to a wide audience of readers.

Lerner Publications Company
241 First Avenue North, Minneapolis, Minnesota 55401